The Gold Brick and the Gold Mine: Fake Mining Schemes that Steal the People's Savings

Emerson Hough

CONTENTS

EDITOR'S NOTE.	I
HONEST MINING IN HANNIBAL'S TIME	1
THE "SUCKER LIST" IN WALL STREET	4
NEW ENGLAND "DONE" BY AN INSANE MAN	6
ALASKA REYNOLDSIZED	7
"ADJOINING" MINES – GOOD BAIT	9
WHEELBARROW VS. $72,000	11
ENTER THE FINANCIAL AGENT	12
CULLED TWO MILLIONS IN FOUR YEARS	13
LYMAN'S SCHEME TO GET STOCKING SAVINGS	16
"SALTING" WITH A CIGARETTE	17
THE MULATOS MINE SALTING SCHEME	18
THE MAIL AND MINING THIEVES	19
POST-OFFICE PROTECTION INADEQUATE	21
LAWS TO PROTECT INVESTORS	24
CAMPAIGN OF THE AMERICAN MINING CONGRESS	26
THE ENGLISH WAY OF MINING – HONEST BUSINESS	28
THE AMERICAN WAY – A GAMBLE	29
THE PROMOTER AND THE CREAM	31
THE PUBLIC = THE MINE	32
CRACKER-BOX INVESTORS	34
ENGLISH VS. AMERICAN MINE REPORTS	36
NEEDED: A FEDERAL BUREAU OF MINES	39
WHAT YOU AND I CAN DO	41

I

EDITOR'S NOTE.

– It is time vigorous efforts were made to stop the cruel frauds perpetrated on the name of one of the world's greatest industries. Mining is a legitimate and honorable enterprise. It contributes immensely to the national wealth. It has been the source of some of our great fortunes. Because there is something magical in the suggestion of gold or coal or copper taken out of the ground, sharpers have made mining an instrument of successful deception. They have tricked people into investing their savings in worthless or even nonexistent mines. Perhaps you who read this have bitten at an advertisement in a reputable publication, which pretended to place the wealth of some western El Dorado at your feet for a few hundred dollars. Doubtless your money has disappeared. It is for the purpose of giving you the protection of a knowledge both of legitimate mining and of the ways of thieves that this article is published.

AMERICA is the land of the free and the country of opportunity for all. Incidentally, it is free hunting-ground for sharpers, and a land of opportunity for the unscrupulous. No such chances for fraudulent business exist anywhere else in the world. Americans are the richest people on earth, and the most easily parted from their money. Those whose sole ambition is to get rich quick very frequently help some other man to get rich quick. Society owes no debt to either of these. It is obliged to support them both. This is wrong both as a moral and as an industrial proposition. Once, a dollar was spent to mine a dollar. today two are spent: One dollar goes into blasting powder, the other into advertising and office furniture.

II

No doubt you have heard the age-old legend of the Mother Vein of Gold, which appears and vanishes, now and again, in this corner of the world. Superstition regarding this great original vein of gold is found wherever men seek the precious metal. The feverish Spaniards called this phantom lode the Madre d'Oro, or "Mother of Gold." Now it is located in Mexico, now in India or Peru, California or Australia. Tradition says that Montezuma got his gold from this great vein, which lay in a secret valley whose whereabouts was jealously guarded by three priests of the war tribe, sole possessors of the knowledge. Any intruder who by chance or design looked down into this valley was smitten absolutely blind. Tradition among the successors of the Aztecs says that when Montezuma passed, the Madre d'Oro sank back again into the earth, and has been seen no more. Men still follow the phantom vein. Those who see it, even in their dreams, still are smitten blind.

Gold! There is no other word that means quite so much. We want gold; indeed, we must have it. Malleable, divisible, indestructible, rare, it is the indispensable medium of exchange. It is our chosen unit of power and success, the measure of civilization and human attainment. Hence it has always been the object of human desire. The Golden Fleece very probably was the sheepskin bottom of an old-time sluice-box, in a day when they used wool, instead of blankets, below the rocker troughs. In the vast ruined civilization of Southeast Africa unknown men once mined probably $400,000,000 worth of gold. There are mines profitably operated in Greece today which the Phoenicians opened 1,200 B. C. Sixteen hundred years later the Romans owned all the mines in Europe. Hannibal once paid his warriors in gold coin of Carthage. Egypt was settled by the Semitic races 2,500 B. C., because of the gold that was found there. A thousand years later Job knew about gold, and five hundred years later still, King Solomon showed what an abundance of wives and what a reputation for wisdom a man can get when he has unlimited gold mines back of him. Columbus found America when he was searching for the wealth of Ormus and of Ind. Cortez and Pizarro toiled and slew in the hope of finding the Madre d'Oro. The great discoveries of the world have been made by men in search of gold. The great voyages of exploration were in part piratical voyages made in search of gold already found and mined by others.

1

HONEST MINING IN HANNIBAL'S TIME

But there is to be said about gold mining ways of the old time, that Tyre sought gold with actual ships, with actual men and mining implements. The peninsula of Sinai did not sell stock, but mined actual gold. Gold in those days meant actual risk and courage. Perhaps even then fraudulent promoters weren't unknown; but he who ventured, in the days of Vespasian or Hannibal or Hiram, too prominently to gild the gold brick certainly lost his head. The mining of gold was then a sober and serious and honest matter.

In America we place the gold brick ahead of the gold mine. We mix alloy of duplicity and greed with the virgin metal of our standard of value. By improved mining methods we nearly double our output of gold, and so cheapen it by well-nigh a half. This shrunken gold dollar is small enough; but that is not all. We adulterate and divide it by, say, another half when we falsely double its cost. This we certainly do when we issue counterfeit promises as against good coin; for in civilization and commerce always the genuine coinage has to pay the cost of the counterfeit. Your tailor charges you a stiff price for your suit of clothes. That covers the clothes of the dead beats who did not pay. To allow the sale of a fraudulent mining stock is to depreciate the basis of this country's values. Such a wrong ought not to be allowed in a country claiming an enlightened government.

It is the thief who is protected in America, not his dupe. The old law of caveat emptor protects the seller of fake mining stocks, not the buyer of them. There is little or no actually enforced law to protect the latter. That is to say, there is little or no actually enforced law to protect those who most need protection, those of small incomes, orphans who have no guardians, wage earners who have little education, widows whose life insurance is not quite enough to support them, women engaged in the desperate battle of life and needing more money, quick money, better to protect themselves. The fence between these and the natural perils of the world is slight enough. In America we break it down entirely.

We offer these helpless ones freely as victims to the greater cunning and strength of men wholly without sense of business honor or personal decency. When we do this, we also attack the whole system of savings banks, which is, or should be, the very bulwark of a nation's financial safety. Says the wolf to the widow, to the busy professional man, to the clerk, the stenographer, the wage earner: "Take your money out of the savings bank. What is three per cent. a year, when I can make you three hundred per cent. a year? Give your money to me!" We permit that. Our national government does not undertake to put a stop to it; our states do not undertake to do so; and this fact is more possible through actual lack of proper statutes than through any misinterpretation or lack of enforcement of the law.

The field is one devised by nature for the trickster. His success does not depend altogether on human gullibility; part of his argument rests on the conditions which surround the industry of mining, one which never can be free of extreme risk. All men know that gold is found far away, where living is high and means of transportation are scarce; that it costs large sums to find and dig it, and that such sums are more easily raised among the many than among the few. None of these attending features has weight to stop the capitalization of bona-fide enterprises. These latter are used as bait by men who have nothing bona-fide to offer, and who make their fattest profits out of their shallowest shafts.

THE "SUCKER LIST" IN WALL STREET

Methods vary among such fraudulent operators, but new victims continually are found. The "sucker list" of one firm in Wall Street numbers 110,000 names, selected as those of persons who will bite more than once at a mining scheme, and whose records show that they have so bitten. This operator proudly declares that the only way a sucker can get his name off that list is to die. In the reorganization of the firm of Douglas, Lacey Co., of New York City, it was discovered that 20,000 persons had money invested in stocks of the company.

The best bait in this particular operation was a "trust fund" established for the benefit of stockholders. The proceeds of the better-paying mines were to be applied to pay dividends for those which were less successful. In this way, the various directors of the many Douglas-Lacey Companies explained, it was impossible for the investors to lose. But they did lose. The reorganization, intended to save some of the better properties, wiped out more than seventy per cent. of the small stockholders – widows, schoolteachers, stenographers, washwomen, scrubwomen – all who once had a dollar in the stocking.

Burr Brothers, Inc., of New York, used the effective bait of the instalment plan of payment. Their literature and advertising offered sudden wealth at twenty cents a share, payments to be in instalments, "the best twenty offers" to be accepted. It was pointed out that if one made one's weekly payment large enough to be included among the fortunate twenty, one could have a nice, clean certificate sent to one immediately, and pay for it at one's leisure. If you think the operators could not afford to do that, you are ignorant. There was an old negro woman in the South who often importuned her white friends for funds to build a certain somewhat mythical church. They asked her what she received for the time spent in collecting. "I has what I gits," was her frank response. She enunciated a great modern mining principle which has made fortunes in Denver, Butte, New York, Boston, and many other places where handsome lithographic work is done, and where advertising space can be bought in journals considered reputable.

NEW ENGLAND "DONE" BY AN INSANE MAN

Sometimes there are victims in enterprises of this sort where there probably was no deliberate intent to deceive or to defraud. Not long ago, in Boston, one Henry D. Reynolds, formerly president of the Reynolds Alaska Development Company, was brought before the United States Circuit Court on the charge of using the United States mails with intent to defraud. Three alienists are said to have declared him insane. In 1907 ex-Governor John G. Brady, of Alaska, endorsed Reynolds and his schemes, and is reported to have collected in New England about $450,000 for these Reynolds projects. Brady gave "lectures" and stereopticon exhibitions in New England churches. Reynolds took out an excursion of Boston and New England investors to Prince William Sound, at one time, and showed them the seacoast of Alaska, practically all of which he claimed to own. At Boulder Bay he took his party into a long tunnel, the face of which they were told was composed of solid copper ore. When they emerged into the garish light of day, each was given a bright copper nugget, said to have come from the mine.

7

ALASKA REYNOLDSIZED

Really, according to local report, these nuggets of native copper had been taken from sluice boxes on Chittitu Creek, 235 miles inland. Reynolds, so ran the story, had treated them with an acid bath to brighten them, knowing that bright bait is better. At any rate, the good, sober New Englanders went back home and sent him $300,000 more, which set him entirely "dippy," in local phrase.

Reynolds's scheme was to run all the barber shops, laundries, bars, and pretty much everything else on the Alaskan coast. A certain Sam Blum had a store and bank; Reynolds wanted it; and Blum, it is alleged, annexed $50,000 of the New England money as a forfeited first payment on his property. A steamship company, it was said, got $75,000 of money on a forfeit. So the good New England savings merrily disappeared, in one of the most spectacular farces ever known in Alaska; which latter is too good and valid and valuable a national possession to permit to be Reynoldsized, as it has been. Reynolds, in the belief of one who knew him well, was a combination of the ignorant enthusiast, the wild promoter, and the crazy man; and as for Brady, another Alaskan called him "nothing worse than an innocent old ninny." Yet, even with so sorry a mental equipment, these two took something like half a million out of conservative New England! The ease with which money can be raised for such enterprises by the deliberately fraudulent or the unintentionally insane continues one of the wonders of our civilization.

Another kind of bait offered is that of the "prominent name." This has proved more useful in England than in this country. Whittaker Wright was able to secure members of the nobility for his boards of directors, and the English public swallowed his schemes one after another, bait, hook, bob, and sinker. In this country we have no lords whom we dearly love, so the names of prominent literary or scientific men sometimes are employed by wise promoters. A "prominent mining expert" is excellent bait. Some good men have been used in this way, and the bait of their reputation in other lines of activity has served to make ignorant and innocent people of small means swallow the hook hid in the lying statements which they have perhaps innocently, certainly ignorantly, fathered. We are all familiar with the literature of this class, sent to us under the guise of personal and intimate confidence. Always that part of the communication is followed by the blackfaced type where the stinger lies concealed. The words AT ONCE usually come in capitals, as do LAST CHANCE, and PRICE POSITIVELY WILL ADVANCE AFTER TEN DAYS. Millions and millions of dollars have been extracted from the public by these means. There is no law against it.

"ADJOINING" MINES – GOOD BAIT

Then there is the same old argument about wonderful properties "adjoining" such and such a dividend-paying property. Very often the properties are miles apart. They might be within twenty-five feet of each other, and one still might be worthless and the other rich. The profits of old and famous properties very frequently are given in advertising literature of this class, "to show what money there is in mining." The "property" sold may be a ten-foot hole in a sandbank two thousand miles from any of these; yet this absurd argument is sufficient to extract coin from the pocket of the American buyer. You can use Michigan to tout him on to Arizona; Utah to land him in California; Mexico to interest him in Alaska. Is it not true? There is no law against it.

Again, the appeal to your mining pocket may come, not through the advertising page, but in the proper person of the promoter or owner himself. For instance, not long ago a gentleman from California came into my office. He owned a mine on the old and well-traced Mother Vein, of Tuolumne County, California. It had been well opened, and showed, in development, according to a reputable engineer's report, three million dollars' worth of ore in sight, with many tons of the best ore already in the dump, stuff which would run very high in value.

At the proper time the gentleman carefully produced from his pocket a little ingot of pure gold, product of one test-mill run. He gave the best of references as to his responsibility. He offered to guarantee ten per cent.

dividends on all money invested, and declared that he had a banking proposition and not a mine.

11

WHEELBARROW VS. $72,000

"My Christian friend," said I to him, "you seem to have a good thing. How far is it from your mine dump to the nearest bank?"

"About five miles," he answered.

"In that case," said I, "it seems to me you don't need to sell a hundred thousand dollars' worth of stock to build a stamp-mill. You need only enough to buy yourself a good, strong wheelbarrow. In two or three months you can thus build your own stamp-mill and pay for it with ore, and still have your mine all in your own hands."

He could not see it that way, and, pursuing his own method, he took $72,000 in two weeks out of the city of Chicago, from some of the best business men of that city. Now, perhaps he had a real mine. I have no right to doubt that he had; but the point of interest to the small investor is this: neither have I any right to believe that he had . The thing for me to do, had I wished to invest in this way, would have been to send an expert to see the property personally.

ENTER THE FINANCIAL AGENT

In this game of plucking the dollars of the poor and the ignorant, there has been a gradual improvement in methods. The constant aim has been, first, to increase the amount of the harvest; second, to reduce to a minimum the risk the reapers run of detection and punishment by the authorities. Experience in most lines of commercial activity has shown that the middlemen often gather in the largest profits and have the smallest losses. Many of those working the mining game - and by this is meant selling stocks on wind and water - have made use of this fact. today in the majority of cases we have, in place of the prospector or the company selling stock direct to the suckers, the financial or fiscal agent. He operates either under the name of a banking firm or as a security company, which is generally a registered trade-name intended as a cloak to cover the names of individuals not desirous of publicity.

The financial agent of this description is in reality the organizer and promoter of the mining company whose stock he sells. But should trouble come along, he is the first to assert that he has been deceived as well as his customers. He sells the shares of the mine on a commission basis so large that practically nothing is left for development. He takes out of the money secured large salaries and the entire expense of advertising and carrying on the exploitation. He prepares all the literature. One of the advantages he claims for his proposition is the wide distribution of the stock as a safeguard against assault by wicked Wall Street interests.

CULLED TWO MILLIONS IN FOUR YEARS

In this wide distribution, however, lies one of his own greatest safeguards against either criminal or civil prosecution. Scattered over the country are his investors – the mill hand, the poor seamstress, the humble artisan, whose total investments, comprising perhaps all their savings, seldom exceed one hundred dollars each; and, with their savings gone, there isn't money left to pay carfare to the office of the financial agent, let alone to undertake a civil suit or enlist the aid of the authorities. The poor seamstress has no way of knowing any of her fellow unfortunates. Hence the utter impossibility of cooperation in seeking to get back their savings.

As an example of the fiscal agent, there may be cited the concern of Douglas, Lacey Company, already mentioned, a concern which in four years, through its operations in this country and in Canada, culled from the people of this country, according to its own statement, over $2,000,000 in exchange for stock certificates in more than forty varieties of mining companies. Here is a letter written to a woman by this concern four years after she had invested all her savings in the stock of one of these companies through this concern, showing the advantage of the fiscal agency plan:

DOUGLAS, LACEY CO.

Financial Agents

66 Broadway. New York

Cable Address "Douglacey" - Anglo-American and

Bedford McNeil Codes

Telephone, 790 791 Rector

DEAR MADAM: June 2, 1908.

Replying to your favor of June 1st would say that we do not find in our files any recent letter from you, and your letter addressed care of 44 Wall Street has probably gone to the Dead Letter Office, from which you will in time receive it.

Now, in reply to your question, we think if you are at all familiar with business procedure, you will see that it would be impossible for the fiscal agents of any of the companies to return money which had been paid for shares and which had been turned over by the fiscal agents to the treasury of the various companies and expended in development work on the different properties.

It is true that we have sold stock for our customers at various times and we are glad to do so when it is possible. At the present time, however, as this company is in process of reorganization, there would be no market for its stock and for this reason we are unable to help you in the way you request.

Very truly yours

Douglas, Lacey Co.

In pursuing this method, few promoters have had the success of Dr. John Grant Lyman. He is credited with having gathered in a half million dollars in his International Zinc operations. This company was supposed to have valuable zinc properties in the Joplin district of Missouri. To unload its stock on the people of this country Lyman organized the firm of Joshua Brown Company, Bankers, incorporated under the laws of West Virginia. Through them the stock was sold until the collapse of the scheme in 1901, when the investors found that what property it did own was heavily mortgaged. While the firm was taking in the money, Lyman maintained a racing stable, had a reputation as a daring automobilist, and even invaded the sacred precincts of the New York Stock Exchange.

LYMAN'S SCHEME TO GET STOCKING SAVINGS

Three years ago the papers throughout this country were filled with the advertisements of the Union Securities Company, selling the stock of the Boston Greenwater Copper Company. It was stated that the mine had cost $200,000 and that so much ore was in sight that an offer of $400,000 had been refused. The Union Securities Company, with offices in New York and in Goldfield, Nevada, started the stock at forty-five cents and lifted it to a dollar. It was merely another name for John Grant Lyman. Not only did the Union Securities company sell the stock to the public, but it also offered it to brokers at thirty-seven and a half cents, on their guarantee that it would not be sold by them at less than forty-five cents. The brokers began getting contracts for the stock and then were told that the Union Securities Company was all sold out.

Shortly thereafter, confederates of Lyman came to these brokers and offered stock to them at fifty cents a share; and the Union Securities Company at the same time telegraphed the brokers that it wanted all the shares it could get at sixty cents. That forced the brokers to buy of confederates; but when they shipped on the stock to the Union Securities Company, expecting to get sixty cents a share for it, Lyman was gone. It had not cost him much. He owed the newspapers of this country $150,000 for advertising, which went unpaid. He reaped $300,000 profits. Boston Greenwater Copper stock can still be found in many a stocking - of humble folk.

"SALTING" WITH A CIGARETTE

It is not, however, always the city promoter who furnishes all of the crookedness. He himself may be deceived by those who sell him the mine. Some of the most thrilling stories in literature might be written about salted mines. The sale of the Bear's Nest Mine, and the special train expedition to the salted Bear River placer field; the sale of the Mulatos Mine to a set of Chinamen, and scores of other instances in American mining history, have been regarded rather as big jokes than as great lessons. And as to such large jesting we advance in finesse. The old way of salting a placer or a quartz vein with a shotgun is now antiquated.

A little while ago a party of capitalists bought a Nevada placer on what they thought to be strictly a "cinch" basis. With their own hands they collected the specimen dirt from all over the claim, and they watched a Mexican miner pan the dirt at the creek. The pans showed up beautifully. They bought the claim. Later, it proved worthless. Afterward they remembered that the Mexican smoked cigarettes all the time he was panning, and that he was careless in expectorating, as well as in knocking the ashes off his cigarettes. The truth was that the highly intelligent Greaser was using the cigarette trick in salting the pan. There was much fine gold in his cigarette and under his lip!

THE MULATOS MINE SALTING SCHEME

All sorts of methods of salting mines, even to the injection, with a hypodermic needle, of strong solutions of mineral salts into a mining engineer's carefully sealed sample bags, have been worked. The most honest, careful, and expert mining engineers have been deceived time and again, and salted right under their own eyes. Even a bland Chinee may be fooled. Take the instance of the Mulatos Mine: The bunch of Chinamen who proposed to buy it insisted on a millrun test on fresh-mined ore, taken out by themselves, for a five-days' run. They were not taking any chances, in their own belief. The owners of the mine, however – so runs the story – had a platform of plank arranged above the timbers at the top of the drift where the Chinamen brought out their ore cars. On this planking a man lay face downward where he could see each ore car that passed. He had a rather hard life for five days on the sandwiches and water which he took up there with him, but he managed to drop a pinch or so of nice gold dust into every car of ore that came trundling under him. The millrun was an entire success from the viewpoint of the sellers, although not from that of the buyers.

There is no working law, let us repeat, which actually protects the investor against this sort of thing, nor which always protects even the promoter, though he be honest. The game is risky all the way along the line, in spite of state laws against the heinous crime of salting, which latter hath as yet by no means lost its savor.

THE MAIL AND MINING THIEVES

As matters stand today, the man selling mining stock on a fraudulent basis fears the Post Office Department much more than he fears the District Attorney. That is the main protection which the public has against such schemes. But to depend upon it is like trying to stop Niagara with a dam of reeds. The man who induces you to take your money out of the savings bank in exchange for stock in a mine, through such operations as have been described, thrives by reason of his use of the United States mails. It is a mail-order business pure and simple.

Let us see what machinery the Government has to protect you and prevent the letter-carrier from bringing daily to your door the flamboyant literature intended to lure your money from the bank. There are five hundred Post-Office inspectors employed in watching Uncle Sam's mail wherever it is carried, in keeping the vast and complicated machinery of the Post Office Department oiled and working smoothly, in running down Post-Office robbers and mail thieves and, lastly, in keeping the mail free from frauds. Ninety per cent. of this force is required to do the routine work of the inspecting branch; that is to keep the machinery running smoothly and to prevent delays. That leaves just ten per cent. for actual detective work such as is necessary in running down thieves and in tracing frauds. In the New York district, which comprises the state of New York as well as New York City, there is a force of twenty-five men working under a chief inspector. Of the ten men assigned to work in New York City, by no means all have special detective ability, and the time of these is taken up almost entirely in catching actual thieves.

POST-OFFICE PROTECTION INADEQUATE

It is only the biggest and most barefaced scheme that under these conditions can receive any attention whatsoever from the department, and even then its force is hopelessly inadequate and incompetent for the work in hand, work requiring the highest-class detective ability.

About twelve years ago the Post Office Department ran down and convicted a swindler, Stephen Balliet, who was selling stock in a mine full of water in Oregon and was known as "the mining genius of the Northwest." He was tried three times, finally convicted, and sent to prison. That case cost the Post Office Department $18,000, took a man's entire time for two years, and required two trips across this continent. The Government has not tried since to get many such convictions.

Perhaps because of the pressure of other work, perhaps for other causes, investigations of this nature are allowed to languish. Some years ago, when the firm of Douglas, Lacey Company was reaping its harvest, an inspector was assigned to investigate the concern's operations. He was one of the ablest inspectors of the service, a man with real detective ability and a knowledge of the devious ways of certain kinds of financing. He made a trip to Mexico and subsequently sent in a report to Washington recommending that a fraud order be issued against the concern and that its use of the mails be stopped. He waited a long time and then got word from Washington that more evidence was required. He made another investigation and sent in another report, recommending in even stronger language that the mails be barred and the public protected. While on this work he was constantly assigned also to other matters and finally was shifted to a station in the South. The concern collapsed some years later, leaving thousands of people in this country and in Canada bereft of their small savings. There was no fraud order ever issued against this firm, though shortly before it closed up it was informed that if it continued to sell stock its use of the mails would be stopped.

The burden of proof is on the buyer. If he turns to the District Attorney he finds perhaps a sympathetic official, without power to assist him. The man selling bogus mining stocks knows all this; therefore his harvest goes on. It is better than the green-goods game, better than the wiretapping swindle, safer than selling any other form of gold bricks. A few years ago a reporter who was engaged in investigating the schemes of Cardenio F. King - now in Charlestown jail, but then posing as "the apostle of the golden rule in finance" and selling his stocks by the barrel in every mill town in New England - made a call on the late John B. Moran, then District Attorney in Boston and widely known as a reformer. He asked Mr. Moran's help in proving that King was a swindler.

"Young man," said Boston's reform District Attorney, "if King was selling corner lots in heaven and advertising them in the newspapers, I couldn't stop him, because I haven't anybody to send up there and prove that they are not there."

King wasn't selling corner lots in heaven, but he was selling stock in a Texas company that was the next thing to it, so far as tangibility is concerned. It was only when he actually took from investors money sent to him to buy real stocks, and pocketed it, that he was put in jail.

LAWS TO PROTECT INVESTORS

A plan for the protection of the investor by statute is embodied in a model law drafted by the American Mining Congress of Denver, and recommended for general passage:

AN ACT.

To Prohibit the Making or Publishing of False or Exaggerated Statements or Publications of or Concerning the Affairs, Pecuniary Condition or Property of Any Corporation, Joint Stock Association, Co-partnership or Individual, Which Said Statements or Publications Are Intended to Give, or Shall Have a Tendency to Give, a Less or Greater Apparent Value to the Shares, Bonds or Property, or Any Part Thereof of Said Corporation, Joint Stock Association, Co-partnership or Individual, Than the Said Shares, Bonds or Property Shall Really and in Fact Possess, and Providing a Penalty Therefor.

Section 1. Any person who knowingly makes or publishes in any way whatever, or permits to be so made or published, any book, prospectus, notice, report, statement, exhibit or other publication of or concerning the affairs, financial condition or property of any corporation, Joint-stock association, co-partnership or individual, which said book, prospectus, notice, report, statement, exhibit or other publication, shall contain any statement which is false or wilfully exaggerated or which is intended to give or which shall have a tendency to give, a less or greater apparent value to the shares, bonds or property of said corporation, joint-stock association, co-partnership or individual, or any part of said shares, bonds or property, than said shares, bonds or property or any part thereof, shall really and in fact possess, shall be deemed guilty of a felony, and upon conviction thereof shall be imprisoned for not more than ten years or fined not more than ten thousand dollars, or shall suffer both said fine and imprisonment.

This law has been enacted in six states and a campaign for its general enactment is under way. But let not the credulous investor suppose that even such a law would guarantee him against loss. The Secretary of the American Mining Congress, Mr. James F. Callbreath, offers the following comment:

CAMPAIGN OF THE AMERICAN MINING CONGRESS

"I do not believe that any one law can effect protection to mining investors, nor that the protection afforded through the Post Office Department forbidding the use of mails for fraudulent advertising matter can fully cover that ground. The greater part of mining frauds are perpetrated without the use of the mails.

"The proposed law, in our judgment, is the longest possible step toward preventing mining frauds. A second step has been taken in the form of a publicity law. My belief is that no system of laws, either state or national, will prevent men from gambling in mines more effectually than such laws now prevent gambling in its more common forms. These may restrict and furnish protection to those who are wise enough to open their eyes, but it will be impossible to protect all the fools all the time. It is the purpose of the American Mining Congress, after having secured the enactment of laws providing penalties for fraudulent representations and requiring publicity, to perfect an organization to secure execution of these laws, and also to carry on campaigns of education showing to investors, first, that mining is a legitimate business and not a gamble; second, that mines are found and not made; third, that investments in mining should be made with the same care and prudence exercised by business men when embarking in other business enterprises. . . . The next work of our organization will be along the line of developing some manner of control of corporations by which paid-up capital stock shall represent actual value."

Mr. Callbreath would seem to be one foredoomed to his own troubles; yet it is clear that he and his organization stand for legitimate mining as opposed to prospect-selling. In strictly accurate phrase, it is the prospect which is found, and the mine which is made and investment cannot properly begin until a body of ore has been blocked out in a proved prospect. Add to the glamor of risk the haze of fraud, and the foregoing will show the nebulous condition of mining investments in relation to mining laws in America today.

What we really need is a Bureau of Mines at Washington. Nobody protects the mining investor. Nobody guards the widest open gate into the savings deposits of this country.

The American Mining Congress, it should be stated, had a quasi pre-inaugural pledge from President Taft in favor of a Federal Bureau of Mines. Toward this we have made a start. A bill establishing this Bureau has already passed both the House and the Senate, and bids fair to become a law. But the activities of this new department will be confined to safeguarding mineworkers. The next step should be to enlarge the province of the Bureau so as to include the supervision of the mining industry for the protection of investors.

It seems quite likely that the states and the nation will need to unite if adequate protection to the investing public is to be expected. But when did state and nation unite to solve a great popular problem? When did section ever unite with section or even resident with nonresident? This is America.

THE ENGLISH WAY OF MINING – HONEST BUSINESS

Back of any movement of this kind there must be popular interest in popular education. Thus far, the greater publicity idea is of more value than anything at hand. We may perhaps. best do our own little part by offering some studies in the theory of mining, showing just why it is risky, and just how we ought to tabulate the risk. In addition to this, we can present, and should perhaps first present, some of the results of intelligent mining as pursued in other countries.

Take the Rand Mines of South Africa, operated on the English basis - mines which turned out more than $12,500,000 in one month not long since. The English method of operating on the Rand is this: A corps of experts is sent to examine a proposed property - that is to say, a proved prospect. If their report be favorable, an estimate is made of the cost of a five-or seven-compartment shaft, to be sunk, say, 3,500 feet. The cost of producing a year's supply of ore for the mill is then considered. The cost of the mill and the cyanide plant is also figured. The total cost is then cast up, and the company is ready to be formed for a half million to five millions of dollars, according to existing conditions. This money is paid in, and is ready to start operations. These men mine carefully, using all possible scientific knowledge and practical experience as guides. The operation may have risk, but it is perforce honest.

THE AMERICAN WAY – A GAMBLE

Now let us examine conditions not infrequent in the United States, by no means assigning wings to all English mining men, or hoofs to all Americans:

A prospector discovers mineralized rock. He locates one or more claims as controlled by the laws of the district where he is. Perhaps others also locate more ground. A little work is done, and then the claims are up for sale. A claim is perhaps sold for a few hundred to several thousand dollars; sometimes the seller receives in addition stock in the company to be formed. No attention is paid to the geology, but a company is formed ostensibly for the purpose of mining, with a capital of one million shares at one dollar par. Perhaps four hundred thousand shares are placed in the treasury to be sold for development purposes. Of course the whole thing is as yet on a wholly gambling basis. The property is still a prospect and not a mine, and hence it is not possible to put it on an investing basis. Comparatively few companies have ever used the services of a real expert, although very possibly the company furnishes a report made from a purchasable local "mining engineer," one of the cheapest commodities in any mining district, where the wide hat and the high-laced boot often take the place of a mining education and a reputable character. This is the stage at which, this is the basis on which, most of the mining "investments" of America are made.

In this state of affairs grafters find their opportunity. Prices in a boom camp are always above any sort of industrial warrant. There were literally millions of dollars poured into Goldfield and Tonopah for claims which never had any careful examination by competent men. Fortunes were made by local promoters and "operators" out of claims which could not show ten feet of actual work. Sometimes the entire capitalization was sold out, and the promoters put the money in their pockets. One operator of this kind sold $130,000 worth of stock, and omitted the precaution of putting even ten per cent. of it in the treasury. Fortunately, he got into the penitentiary. Many of his fellows never had actions brought against them except under the postal laws, which naturally are inefficient. There was one shaft of a hundred feet which cost twelve thousand dollars, charged up to the stockholders, the names of dead men being used on the pay rolls as "laborers." The mine boss and the local officers got big salaries to keep their mouths shut. The real mine was in the savings banks of America, in the pockets of non-residents. In Nevada alone, in the past four years, more than twenty million dollars have been invested in worthless properties. One engineer with a government certificate could have saved the clerks, stenographers, widows, washwomen, and orphans of America fifteen million dollars at the cost of, say, five thousand. Would that have been a good investment? What could a dozen do? What could an efficient corps do? Is there here yet one more future task for our patient and longsuffering United States Army? What police work would pay better dividends?

THE PROMOTER AND THE CREAM

Even when the mine wins, the small stockholder rarely wins. The promoters often take the cream. Suppose a company is organized for three million shares. One million is put in the treasury for sale. Of this million shares, say, two hundred thousand are offered at twenty-five cents. This raises a working capital of fifty thousand dollars. Let us be very glowing, and suppose that, with this fifty thousand dollars, we really uncover five million dollars' worth of ore. The net profit would not exceed three million dollars; so that the man who put in twenty-five cents might, after a long time, get back a dollar. In the meantime, two million dollars would have gone to promoters, in "commissions," and so forth. There are thousands of such cases, and still the people continue to bite on such bait.

THE PUBLIC = THE MINE

Instances of actual Nipissing rises caught in time by the lamb are very rare. I rom first to last, the public is the mine, and the returns come out of the savings banks. In some mines "high grading" – the carrying away of valuable pieces of ore by the miners themselves – is fought as sternly as the diamond stealing by the Kaffirs in a Kimberley mine. In yet other mines, far more numerous, high grading is encouraged among the miners. The report gets out that the ore is so rich that the miners steal it in their dinner pails. That booms the stock. Wall Street makes this money out of the market and not out of the mine.

In spite of all warning and all examples, the average American will to a certain extent persist in gambling in mining stocks. Supposing this to be true, it is of value for the investor to learn something of the theory of mines, something enabling him to pass on the natural value of any mining stock which is offered to him. What, then, is a mine? What are some of the inevitable features in developing a mine?

In the first place, there must be prospecting. This is sheer and unavoidable risk on the face of it, and it is attended with economic waste which cannot be avoided. Of a hundred prospectors, ninety-nine die poor. The failures must be charged off to industrial waste attendant upon inherent conditions of the mining industry.

Again, in the development of a mine after it is located and proved in part, there is more unavoidable economic waste. The rock is blank and silent. It can only be explored by means of expensive drifts and drillings. In one mine at Bisbee, Arizona, a shaft was sunk which had drifts at the 600-and 900-feet levels, all without result. Later on they found a blanket of copper between those two levels, from which six million dollars were taken. Even in old established mines there is something of a chance, and there are often unwittingly false standards of values. Which is no argument for making all gamble that which originally was part gamble.

Any mine, no matter how rich, or how large, begins to be exhausted from the time the first pick is stuck into the ground and all its profits ought to be figured on the basis of diminishing deposits. When your deposit is drawn out, your bank does not honor your check. A mine is the reverse of a mortgage or a bond. The security does not remain stable nor increase in value, but, on the contrary, continually decreases in value. In a mortgage, six per cent. is wisdom; in a mining return, it is folly. A mine, instead of being figured on the basis of a mortgage, ought to be figured on the basis of a term annuity. That is to say, on the basis of a wiping out date. When the mine is done paying dividends, there is no return of the face of the principal invested. Yet the great and gullible public forgets this all-important fact, which differentiates mining from every other form of business.

CRACKER-BOX INVESTORS

There is every probability that the average investor never heard of a proper "amortization charge" in the management of a mine. Until he shall have heard of it, until he shall have learned something of the terms of life annuities, he ought never to invest a cent in any mining stock. After he actually has learned the theory of amortization, he will observe that almost every mining stock listed in public prints is selling at an inflated value. That is to say, even the best and most stable of mines are overrated, not to mention the purely wildcat ventures. Some mines may naturally be long-lived, others short-lived; yet, if either pays a good, stiff dividend, the public makes no distinction between the two and will buy the stock of either. In this investing, the public has no protection on the part of the government, on the part of honest publicity, or on the part of its own careful education.

In the majority of cases, a mine ought to pay annually perhaps twenty per cent. of the investment, to be profitable. That is to say, the actual value of any mine is rarely over five times actual dividends paid after expenses of operation. How many mines are capitalized on any such real basis as that? The answer lies in our own ignorance, and in the shrewdness of the men who sell us mining stocks. Stocks that are the best dividend-payers often sell at ten or twelve times the face of the annual dividends. Let the mine hit a brief streak of bonanza, and the stocks will climb yet higher. We buy such stocks, or worse; but even a fundamental acquaintance with the theory of mines would show us that such an investment is usually a bad one. In a mortgage we do not look to the interest to pay us back our principal; in a mine we must look to dividends to pay us back our principal and interest also. When the mine is done, our principal is gone. But how many mining investors ever thought of that? And how many, when offered a ten per cent. "guaranteed dividend" for five years on their money, ever stop to reflect that, for instance, I could take your money and put it in a cracker box, and myself make money by paying it back to you, ten per cent. a year for nine years - and then explaining what had happened to the cracker box! Now, most of us are just such cracker-box investors. We pay out millions and millions annually, just that foolishly. And our nation, our states, allow us to do it. They even - as recent legal proceedings prove - allow the "inside" operating stockholders to borrow money to pay dividends to the "outsiders." That keeps up the "values" in the market. It does not enhance the real value in the mine.

ENGLISH VS. AMERICAN MINE REPORTS

Again, granted even a valid and a well-managed mine, how much information regarding it does the average investor in the stock secure? In a general way, he knows in advance that all mining, whether placer or quartz, is very expensive. Beyond that, he gets the annual report of the officers, which will tell perhaps the names of the men who are spending his money, the total earnings, the total output, the balance sheet, the statement of capital stock issued - and little else. All of which means nothing!

A well-regulated English company is obliged to go much farther than this. A good annual report will show the advertisement of the general meeting of stockholders, the list of directors and officers, reports of directors, giving details of the condition of property, including the development work, the tonnage of production, the values recovered from such tonnage, the costs of operation, the profits for the period covered, the balance sheet of accounts, the profit and loss statement, including a working cost estimate, the appropriation list showing what has been done with all the earnings, the reports of managers giving details of the development work, the estimated values of ores exposed on three sides, the probable values of ores not so well exposed, the working expenses, the construction account, general remarks on the physical condition of the property, and a map of the property itself.

What American promoter would trouble himself to make such a showing as that to the American sucker? Even if such detailed information existed in the records of the average American mining concern, the sucker could not get access to the books even did he have the temerity to demand it.

Professor H. S. Munroe, of the Columbia School of Mines, when asked whether such a thing as general supervision of mining investment could be possible, answered: "Yes, if some philanthropist will give us ten millions to endow such an institution, and maintain a corps of engineers in the field who will do work similar to that accomplished by J. Curle under the auspices of the London Economist. Such work should, of course, cover all incorporated mining companies, not merely a few hundred of the more prominent gold mines; and it should be continuous and not spasmodic. Such a plan is of course Utopian, but I feel that anything less would be likely to do little good. Even Curle's opinions began to lose their value within a month or two after they were written, and are of less value every year. Mining can never be put on the same basis as agriculture, for the reason that the risk of failure is infinitely greater, and that it is impossible to prove the value of any mine or mining region without spending a large amount of capital, the greater part of which will inevitably be lost in this work of initial development."

Those are the sober words of an expert who spends his life in studying the theory and practice of mining. If such words shall teach us a little wisdom, so much the less need for laws. But let us consider what the laws ought to do in order to protect you for the sake of your family, and for the sake of society, and for the sake of the savings which lie back of the prosperity of this country.

Let us agree that no government can guarantee the safety of any investment. Let us admit that digging gold can never be put on the same amortization basis with digging potatoes, for instance, because the soil remains for more potatoes, whereas the ore of a mine is exhausted and does not raise more ore. Nevertheless, although the industries of potato growing and ore digging are not the same, the principles lying back of them ought to be precisely the same; and our governments, both state and national, ought to see to it that they are kept precisely the same, and controlled on the same plane legally. If it be true that no government can watch after every mine, none the less any enlightened government can establish general conditions for engaging in mining or engaging in the sale of mining stock; and, perhaps with yet better results, it can establish a general supervision over the mining intelligence of the public, just as it does over the agricultural intelligence of that public.

NEEDED: A FEDERAL BUREAU OF MINES

The enactment of good mining laws, punishing the proved intent to commit a fraud as well as the fraud itself, and seeing to it that capital stock shall be paid up, seeing to it also that all moneys spent by a mining corporation shall be traceable from start to finish, is the natural first step toward the purification of American mining methods. Beyond that, the national government could take a hand in the game through a federal Bureau of Mines. There must be some clearinghouse of intelligence and of values in this country, some place from which our intelligence may start and to which it may return. The public must have accessible reports of engineers, state or federal, of a sort entitled to confidence.

The nest of vermin in our large cities, inhabited by those who make a living out of the ignorance and eagerness of small investors, must be smoked out once and for all. In this work, state and national governments, popular education and intelligence, and the aid of the better class journalism of America, all must be enlisted. The pages of our press might well be far cleaner than they are. The publication which prints the advertisements of a fake-mining enterprise is itself a party to the fraud. A Bureau of Mines chief can sit behind the desk of every advertising manager in the counting-rooms of every newspaper and magazine in America. The press of this country, when it likes, can, by taking thought, somewhat dim the splendor of the mahogany in many an elegant suite of offices in New York, Boston, or elsewhere. It can reduce the reckless and senseless expenditure of ill-gained wealth which is making civilization a mockery in America, and branding our republican form of government as a failure.

We will have a different way of life, or another form of government. We will have a better administration of law in the United States or we will have another political party, possibly another political system. We will clear up this rotten society, or we will try how we like a different organization of society. The people of America are beginning to murmur. The burden of the murmur is that they have long enough been betrayed. Unspeakable injustice has been done the people of America under the forms of law and government. It is coming to be said that our law and government have not an even hand for all, that a few are allowed to despoil the many. When a people murmurs, let a government beware. Meantime the more that certain unspeakable things are reduced in, and eliminated from, Wall Street and the other "financial centers," the better for our schools, our taxes, our farming, our industry, our living, our character, our country.

After all, the government of this country, as we now have it organized, depends on the character of its average individual citizen. The end of this abuse of fake-mining enterprises begins now, here, with you and me, in our intelligence, in our love of a square game. By taking thought we can add a cubit to our own stature, and so add to the stature of our laws and of our national morality.

WHAT YOU AND I CAN DO

As for you and me, when next we see the flaming advertisement advising us that the Madre d'Oro, Montezuma's fabled Mother Vein of Gold, has once more come to the surface of the earth on Manhattan Island or near Plymouth Rock; when next we read counsel that because mining pays in Michigan it ought to pay in Nevada; when next we are advised to get into the game at once because this is our LAST CHANCE - we might at least ask to see the report of the engineer, likewise the record and antecedents of the engineer; and many, many other things. Perchance we might write and ask the mining promoter what, in his belief, is the proper amortization charge in his particular mine. At which the average mining promoter would probably fall dead.

www.ReadHowYouWant.com

You can buy our **Large Type** and **EasyRead** books from our www.ReadHowYouWant.com website, from websites like Amazon.com and through your UK and North American bookshop.

EasyRead books are designed to make your reading easy and enjoyable. **EasyRead** books are published in different font sizes, so you can select the font size best for you.

EasyRead is for people with normal eyesight who want books in an easy-to-read format.

EasyRead Comfort is for people who find reading small print tiring but do not need large print.

EasyRead Large is for people who find it easier to read larger print.

The **EasyRead** font, character, word and line spacing have all been set to make it as easy as possible both to recognize a word, and to run your eye along a line of text without losing your place. We split as few words as possible at line ends, as split words make reading harder and can be annoying. For out-of-copyright books, words have been changed to modern spelling (where the sense is not affected), and the original hyphenation of compound words has been retained where this improves word recognition or sense.

With most things you buy, you get a choice, and you can choose the make and model that suits best. There is a big exception – books. You don't get to choose a format that is easy for you to read – you have to read the format the publisher selects.

This **"One Size Fits All"** paradigm **no longer** need apply to books.

At www.ReadHowYouWant.com, you can order a book just as you want it, so you can read it easily. You can choose nearly any format, and at a surprisingly low cost, because of a technical breakthrough that allows us to typeset an individual book automatically, print and bind the book and have it quickly sent directly to you.

You can have your book in **Talking Book** formats (**DAISY or MP3**) or as an **E-book** or in **Braille**.

We have developed totally new visual formats which we hope will help people with **dyslexia** and other print reading disabilities. Look at www.ReadHowYouWant.com for more information.

Made in the USA
Monee, IL
04 November 2026

67375605R00030